Organic Production of Some Agronomic Crops

Basmati Rice, Cotton, Red Gram, and Sugarcane

Gowri Vijayan

All Rights Reserved. No parts of this publication may be reproduced, stored in a retrieval system, or transmitted, in any form or by any means, electronic, mechanical, photocopying, recording, or otherwise, without the prior permission of agrihortico

© 2014 AGRIHORTICO

Preface

This small book is a sincere attempt to provide some basic insights into organic crop production practices and organic certification procedures, especially for those who have no idea about the topic of organic farming. Techniques mentioned in this book are strictly in an Indian context. Production techniques and all other information had been written keeping in mind the present scenario of Indian organic market and crop production practices. However, this book may be used as a reference material for all situations pertaining to organic production technology.

TABLE OF CONTENTS

BASMATI RICE (*ORYZA SATIVA*) 7

- INTRODUCTION 7
- CLIMATE 7
- SOIL 7
- PROPAGATION 7
- VARIETIES 7
- SEED RATE 7
- SEED TREATMENT 8
- NURSERY 8
  - *Wet method* *8*
  - *Dry method* *8*
- AGE OF SEEDLINGS 8
- SPACING 8
- LAND PREPARATION 9
- PLANTING 9
- MANURING 9
- IRRIGATION 9
- WEEDING 9
- PEST MANAGEMENT 9
  - *Brown plant leafhopper (Nilaparvata lugens)* *9*
  - *Mealybug (Brevennia rehi)* *10*
  - *Gall midge (Orseolia oryzae)* *11*
- DISEASE MANAGEMENT 12
  - *Rice blast (Pyricularia oryzae)* *12*
- HARVESTING 12
- YIELD 12

COTTON (*GOSSYPIUM* SP.) 13

- INTRODUCTION 13
- SELECTION OF SITE 13
- VARIETAL SELECTION 13
- FIELD PREPARATION 13
- SEED RATE AND SOWING 13
- MANURING 14
  - *Green manure and green leaf manure* *14*
  - *Crop rotation and intercropping* *14*
- CULTURAL OPERATIONS 15
  - *Weed management* *15*
  - *Water requirements* *15*
- PEST MANAGEMENT 16
- DISEASE MANAGEMENT 17
- YIELD 17

RED GRAM (CAJANUS CAJAN) 18

- Introduction 18
- Soil and climate requirements 18
- Planting season 18
- Intercropping and Mix cropping 18
- Varieties 19
- Seed selection and treatment 19
- Seed rate and sowing 19
- Field preparation 19
- Cultivation 20
- Weed management 20
- Green manuring 20
- Water management 20
- Pest management 20
- Disease management 21
- Harvesting 21
- Yield 21

SUGARCANE (SACCHARUM OFFICINARUM) 22

- Introduction 22
- Climate and soil 22
- Varieties 22
- Land preparation 22
- Spacing 22
- Organic manure 22
- Planting material 22
- Seed-sett treatment 23
- Sett rate and planting 23
- Green manure intercrop 23
- Crop rotation 23
- Weed management 23
- Biofertilizers 23
- Trashing 24
- Pest management 24
  - *Early shoot borer* 24
  - *Inter node stem borer* 24
- Disease management 25
  - *Red rot disease* 25
  - *Smut disease* 25
  - *Grassy shoot disease* 25
- Cane harvest and yield 25

Basmati Rice (*Oryza sativa*)

Introduction
Basmati rice is known as the King of Rice and is priced for its characteristic long-grain, subtle aroma and delicious taste. Punjab, Haryana and Western Uttar Pradesh are traditional basmati rice growing areas. The photo-insensitivity of semi dwarf basmati varieties like Pusa Basmati I, allows its cultivation in any part of India.

Climate
The crop requires 20°-35°C temperature throughout its growing season, clear skies, low night temperature and a fair rainfall distribution.

Soil
Heavy, neutral soils like clay, clay loam and loamy are preferred soil types for growing basmati. The crop prefers soil with pH range of 5.0 - 8.0. Alkaline and saline soils are not suitable for the crop.

Propagation
Rice propagation is through transplanting of seedlings from nurseries.

Varieties
While selecting varieties for cultivation, remember to select varieties naturally resistant to pests and diseases of that area. The variety should also be suited for the soil and climate of the area. Two of the most popular export quality varieties cultivated are Basmati 370 and Taraori Basmati. Pusa Basmati I, World's first high yielding, semi dwarf variety of Basmati has also been developed. Basmati farmers of Punjab, Haryana and Uttar Pradesh use Pusa-1121, CSR-30. The grain from 1121 has better aroma and lower chalkiness than Pusa Basmati-1, while yields are more or less the same. CSR-30 has higher yields and is saline resistant compared to the traditional cultivars.

Seed rate
Transplanting: 60-85 kg/ha

Broadcasting: 80-100 kg/ha

Dibbling: 80-90 kg/ha

Seed treatment
Dry seed treatment: Dust seeds with talc based formulation of Psuedomonas fluorescens (P1 and P14) at 10g/kg seeds, during sowing time

Wet seed treatment: Soaking seeds for 12-16 hours in solution of P.fluorescens (P1 and P14) prepared at 10g/liters of water/kg of seed

Nursery
There are two methods of raising seedlings in nursery, essentially based on water availability:

Wet method
Raise beds of 5-10 cm height, 1-1.5 m width and of convenient length with drainage channels between beds. The total seeds beds area should be 1000 m^2/ha of field to be transplanted in. Vermicompost at 500g/m^2 and rice husk ash at 100g/m^2 are added to the field while preparing beds. Thrips incidence can be reduced by application of vermicompost. Vermicompost can be substituted with compost or cattle manure at 1kg/m^2, if unavailable.

Dry method
Raise beds of 1-1.5 m width and 15 cm height and convenient length. Apply vermicompost at 500g/m^2 and rice husk ash at 100g/m^2 into the soil while bed preparation. Sow seeds, dry seed treated as described above, evenly on the bed and cover with fine soil/sand.

Age of seedlings
Seedlings are ready to be removed, when they attain the 4-5 leaves stage. The seedbeds should be irrigated lightly a day before pulling out seedlings. Wash off mud and soil from the roots and bundle seedlings based on size.

Spacing
General spacing recommended is 30 cm x 25 cm for rice

Land preparation

The main field is dry ploughed three weeks prior to planting and then water-logged with 5-10 cm standing water. Organic manure (10t) or green manure (10-20t) is incorporated into the soil and properly levelled. The field should be again flooded at least 3 days before transplanting of seedlings. Bio-fertilizers like Azospirillum or PSB/PSM at 2-3 kg/ha mixed with 25 kg FYM or vermicompost can be applied to soil just before planting. In case of very acidic soils, liming is essential (pH less than 5.5). Apply lime at 600kg/ha in two split doses, 350 kg/ha as basal dressing and 250 kg/ha as top dressing about a month after transplanting.

Planting

Three week old seedlings from the nursery are transplanted into the puddled field, based on a rectangular grid.

Manuring

Blue green algae at 10kg/ha can be applied 10 days after planting. Azolla can be applied at 1t/ha, 7-10 days after transplanting and incorporated after 3 weeks.

Irrigation

Water management is very essential in this case, to encourage better nitrogen uptake and good tillering. The field should be levelled properly and drainage system management efficiently. Throughout the growing season, it is recommended to manage a water level of 2-5 cm.

Weeding

The water logged condition of field prevents extreme spread of weeds. Hand weeding is practices 2-3 times in 20 days interval, starting from the third week of planting.

Pest management

Brown plant hopper, gall midge, yellow stem borer are some of the common pests seen on high yielding varieties of rice. Using natural enemies of pests is a great way for organic rice production.

Brown plant leafhopper (*Nilaparvata lugens*)
-Use of resistant varieties

-Close planting should be avoided and a rough spacing of 30 cm for every 2.5 m is to be maintained, to reduce pest spread.

-Intermittent draining should be practiced

-Use of light traps to monitor the pest population

-Release of natural enemies like *Lycosa pseudoannulata, Cyrtorhinus lividipennis*

-Neem kernel 5% spray at the rate of 25kg/ha or neem oil 2% at the rate of 10Lts/ha.

Mealy bug (Brevennia rehi)
-Removal of grasses from bunds and trimming of bunds before transplantation

-Destruction of infected plants

-*Adelencyrtus* sp., *Dolihoceros* sp., *Gyranusa* sp., *Parasyrphophagus* sp., and *Xanthoencyrtus* sp. parasitize on mealy bugs

-Use of predators like *Gitonides perspicax, Leucopis luteicomis, Scymnus* sp., *Pullus* sp., *Anatrichuspygaeus, Mepachymerusensifer.*

Paddy stem borer (*Scirpophage incertulas*)

-Use of resistant varieties

-Clipping of tips of seedlings before transplanting to avoid egg masses

-Avoid close planting

-Destroy egg masses on collection

-Use of light traps to monitor pest population

-Use of *Trichogramma japonicum* and *T.chilonis* parasite 5-9 times at 10adults/m² and one release site per 100m² at an interval of 7-10 days, decrease pest damage by 60%.

-Application of *Bacillus thuringiensis*

-Use of neem seed kernel 1%

-Pheromone traps installation at rate of 20/ha with 5mg impregnate

Gall midge (Orseolia oryzae)
-Early planting encourages avoiding infestation

-Use of resistant varieties

-Proper harvesting and post-harvest field management

-Use of fast growing varieties

-Light trap setting at rate of 1/ha for monitoring purpose

-Use of infrared light trap

-Use of larval parasitoid Platygaster oryzae at the rate of 1 per 10m^2 in the main field, 10 days after transplanting

-Introduction of predators like carabid beetle (*Ophionia indica*), spider species like *Tetragnatha* and *Argiope catenulate*

Leaf folder/ Leaf roller (*Cnaphalocrocis medinalis*)

-Use of resistant varieties

-Clipping of affected leaves to reduce pest population

-Maintenance of bunds from weeds and grasses

-Setting up of light traps to attack and kill the moths

-Use of *Trichogramma.chilonis*

-Neem seed kernel extract 5% spray at the rate of 25kg/ha

Some of the common pest management practices followed are given below

-Calotropis branches should be placed in the irrigation channel. The alkaloid in latex of the plant will act as insect repellent.

-Broadcasting of custard apple (*Annona squamosa*) leaves or seeds in the field. The smell will repel insects.

-Beating heavy drums or standing scarecrows in fields to frighten birds off the crop

Disease management

Neck blast disease, Sheath blight, False Smut, Sheath rot are some of the major diseases that take a heavy toll on the rice crop. The application of bio control agents like *Trichoderma viride* and *T.harzianum* helps control blast disease.

Rice blast (Pyricularia oryzae)

-Application of crushed bark of *Careya arborea* (2-3 kg) to field

-The concentrate of boiled tulsi leaves (1 kg in 2 liters of water) is sprayed at 15 days interval (twice) at the rate of 2ml/liter of water.

Some of the common disease management practices followed:

-1 liter cow's urine+1 liter buttermilk+8 liters of water. Spray this extract over crop to control bacterial and fungal diseases.

-300 ml Sweet flag+1 liter cow's urine+8.7 liters of water. Spraying of solution helps control the diseases.

Harvesting

Harvesting is undertaken as soon as the grains mature, in order to avoid shattering of grain and development of sun cracks. A standard 25-30 days after flowering, for early and medium duration varieties and 35-40 days after flowering for late varieties is considered as proper stage of harvesting. The moisture content should be about 20% during harvest. The harvested grains are then sun dried to reduce moisture to 14% for further processing and storage.

Yield

The yield of the crop reduces during the conversion period because of non-use of chemical fertilizers. However, the yield stabilizes to 90% of normal level by the 4th year of cultivation.

Cotton (*Gossypium* sp.)

Introduction
Cotton plays a dominant role in the Indian agricultural and industrial economy. One of the most important fiber crops, cotton is the very base of textile industry (70%), and contributes a major chunk to agri. export from India (38%). There are four cultivated species of cotton namely, *Gossypium hirsutum, G. arboretum, G. herbaceum* and *G. barbadense*. India is the only country where all the four species are grown. Some of the major cotton growing states in India are Maharashtra, Madhya Pradesh, Gujarat, Andhra Pradesh, Karnataka, Tamil Nadu, Punjab, Haryana and Rajasthan.

Selection of site
Fields free from perennial weed infestation and extreme soil erosion should be used for organic cotton cultivation. Black cotton soils, with an average water content of 100-500mm/m, pH from 7.0-8.2 are ideal for cotton cultivation.

Varietal selection
High yielding varieties may not suit the specifications of organic farming. Hence varieties, disease resistance, hardy and pest tolerant should be selected. Such proactive selections will help in management of pest and disease incidences, especially in absence of chemical control measures.

Field preparation
Deep ploughing every 3 years and shallow ploughing twice a year during summer, are essential for cotton crop. This helps to manage deep rooted weeds and destroy soil borne pathogens and pests.

Seed rate and sowing
Use of acid delinted seeds is prohibited in organic farming. But farmers often use them to reduce expenditure and increase profits. If acid delinted seeds are used for cultivation, the cotton fiber cannot be labeled as 'organic'. A seed rate of 25kg/ha at 75cm x 15 cm spacing ensures 85,000-90,000 plants/ha. Fodder cowpea (*Vigna unguiculata*) is to be planted at the rate of 1:2 rows and ploughed down and mixed with soil before flowering. Seed inoculation with *Azotobacter* or *Azospirillum* (200g/seed requirement

per acre) is recommended. The seeds could also be treated to a mixture of beejamrut (200g/kg of seeds) and *Trichoderma viride* (8g/kg of seed). The treated seeds should be kept in shade and later treated with *Azotobacter* and PSB (5 g each/kg seed) and shade dried. The seeds should be sown within 6-8 hours of treatment.

Manuring

Addition of FYM (15t/ha) to soil during land preparation is recommended. Treatment of FYM with *Trichoderma viride* before application to soil is advised as a preventive measure. Vermicompost (1-2 t/ha) can be used to supplement FYM on the furrow lines after sowing.

Green manure and green leaf manure

Diancha (*Sesbania aculeate*) is raised at 2m width around cotton fields, the lopping of which is then added to the crop rows 65-70 days after sowing. It is also a temporary mulch and source of nitrogen, during the early boll development period of the crop. The growing and burying of fodder cowpea into the soil at 40 days after sowing ensures nitrogen availability to the crop during its high growth phase and flowering period. Weed management, increase in microbial activity in the soil, natural enemy buildup are among a few of the benefits of green manuring. It also helps to control soil erosion to an extent.

Neem, babul, pongamia, sesban, *glyricidia* etc. can be planted as green leaf manure sources. Along with being nutrient sources, they will also act as bird attractants, for pest control.

Crop rotation and intercropping

Crop rotation with a legume is recommended. Cotton is commonly grown in rotations of:

-Cotton-wheat-sugarcane-cotton

-Cotton-red gram-sorghum

Intercropping of cotton with sorghum and red gram is the most commonly followed combination. Intercropping with crops like green gram, black gram, soybean with cotton is also picking up. For optimal output, reduces pest incidence, maintenance of soil fertility, the following intercropping combinations are recommended:

- 1 row maize/sorghum+2 rows of red gram+4 rows of cotton+2 rows of cowpea/soybean+4 rows of cotton+2 rows of red gram+1 row of maize/sorghum

- 4 rows of cotton+2 rows of cowpea/soybean+4 rows of cotton+1 row of mixed plants of red gram, maize and sorghum

- 1 row of marigold/hibiscus for every 20 rows of cotton

- 1 row of maize/sorghum+4 rows of cotton+2 rows of red gram+4 rows of cotton+2 rows of red gram+4 rows of cotton+1 row of maize/sorghum

- 100 marigold/hibiscus plants at random per acre etc.

Cultural operations

Pruning of the shoot tips encourage tertiary branch growth, with many flowers and bolls. Increase in productivity up to 30% has been recorded through proper pruning.

Weed management

The first 2 months after sowing is crucial for cotton growth. Regular weeding at 25 days and 55 days after sowing is required. Mulching could be practiced to suppress weed growth. Sanitation of the field after weeding is necessary to prevent further occurrence.

Water requirements

Cotton requires very less water during its early stages of growth. Water requirement is essential during flowering and boll formation stage. The soil moisture should never go below 50% at any time. Black soil cultivation requires protective irrigation every 20 days in case of monsoon failure. Irrigation by furrow and alternate furrow method are most effective for cotton crops. Drip irrigation is also suitable. The use of mulches and other measures to reduce evaporation is however, recommended under rain-fed conditions.

Pest management

Crop protection measures for organic cotton cultivation generally involves use of predators like *Chrysoperla* sp. or *Apertochrysa* sp., egg parasitoids like *Trichogramma*, larval parasitoids like *Habrobracon* sp., Nuclear Polyhydrosis Virus (NPV), *Bacillus thuringiensis* var.*kurstaki*, utilization of bird perches and botanical insecticides like neem products.

Some of the common pest suppression strategies recommended for organic cotton cultivation are:

-Selection of pest resistant varieties

-Depotting of older cotton plants (> 80 days). This will reduce egg laying by *Heliothis armigera*

-Use of pheromone traps (10/ha) for monitoring for bollworm population

-Spray of neem seed kernel extract 5% or neem seed oil 1%

-Attracting predatory birds using bird perches and other methods

-Use of *Habrobracon hebator* for controlling bollworm larvae and other caterpillars

-Release Trichocards (5/ha) at 50 days, then twice every 12 days, in order to kill bollworm eggs

-Spread *Chrysoperla* sp. (500-100/ha) at 25 days of crop growth, for jassid control

-Spraying of 3-4 weeks old, fermented buttermilk solution (300ml/15liters of water) to control bollworms, caterpillars and spider mites

- Flour spray (2 cups flour+1/2 cup soap water) and soft soap spray (15 g soft soap powder/15 liters of water) are found effective against aphids, jassids, spider mites, thrips and white fly

-Setting up of yellow sticky traps (10/ha) is effective against pests especially white fly

Disease management

Root rot, wilt and browning of leaves are common diseases seen in cotton. Some of the general preventive and control measures are given below:

-Deep ploughing during summer

-Seed treatment with *Trichoderma viride* will control occurrence of root rot and *Fusarium* wilt

-Use of neem products are effective against soil borne pathogens

-Rust and root rot management through spray of fermented buttermilk (5 lts) in lime water (100 lts) per hectare

-Foliar application of *Trichoderma viride* powder (25 g) + milk (50 ml) + water (10 lts) reduce brown leaf patch incidences

Yield

The average yield of an organic cotton crop is 8-10 q/ha under rain-fed conditions and 20-25 q/ha under irrigated conditions.

Red gram (Cajanus cajan)

Introduction
Red gram, also known as pigeon pea (arhar or tur) is a very important pulse crop for India. Maharashtra leads the country in its production. The nitrogen fixing capability of the crop makes it an ideal choice as intercrop. Both monocropping and intercropping systems are followed for red gram cultivation in the country.

Soil and climate requirements
Red gram requires well drained, medium to heavy soils of pH range from 5-7 for growth. Soils with waterlogging nature, high saline content are unsuitable for red gram cultivation. The critical flowering and pod formation phase of the crop requires a temperature of more than 30ºC and low moisture. General temperature requirement is less than 25ºC.

Planting season
Red gram is grown during the monsoon months (June and July). The ideal time for sowing is from mid-week of June to mid-week of July.

Intercropping and Mix cropping
The nitrogen fixing nature of the crop makes it an ideal intercrop. Sorghum, pearl millet, maize, sugarcane, soybean/cowpea or cotton are usually intercropped with red gram. A widely practiced arrangement in cotton is the 8:2 rows of growing cotton and red gram respectively. It is recommended to intercrop red gram with moong and soybean/cowpea during 1st year of field conversion to organic. It has shown significant raise in soil nutrient quality. Though monocropping in red gram is practiced, it is generally avoided because of increased chances of pest and disease infestation. In order to control this, mixed planting of red gram seeds with sorghum seeds (1-2%) or any other millet is recommended. A 2:2 rows of moong and red gram respectively has been recognized beneficial. Random planting of marigold planting (100/acre) also help in pest and insect control.

Varieties

Selection of varieties should be done considering the soil type, inputs availability and local conditions. In case of farming dependent of rains, selection of short duration varieties is logical. For organic farming, it is also suitable to select a hardy/disease/pest tolerant variety. The market demand for the type of grain also has to be seen before final decision.

Seed selection and treatment

Organic certified mother stock should be used for seed procurement. It can be wither obtained from an organic certified farm or can be taken from self-plot. In case of self-plot, care should be taken to select healthy, disease and pest free plants for harvesting seeds. Tagging of the plants selected will help in separate harvesting of the seeds. Drying up of seeds after harvesting should be done to bring down moisture level to 8%.

A mixture of Trichoderma viride (8g/kg seed) and beejamrut (200g/kg seed) is to be made. The seeds should be shade dried after treatment with the above specified mixture. The seeds should then be treated with Rhizobium and PSB biofertilizer (5 g each/kg seed) and again shade dried. Thus treated seeds should be sown with a time gap of 4-6 hours of treatment.

Seed rate and sowing

Seeds are sown by drilling (4-6 cm depth) on wet soil. Crop variety and duration determine the spacing and quantity used. Long duration varieties needs 12-15 kg seeds/ha spaced at 60cm x 20 cm. Medium duration varieties follow spacing of 60cm x 20 cm at rate of 15 kg seeds/ha. Early maturity varieties require 20kg seeds/ha to fill a spacing of 120cm x 30 cm.

Field preparation

Red gram being a deep rooted crop, requires loosening of soil for root penetration. Tillage operations thereby require one deep tillage up to 1.5 feet, followed by a shallow tilling. A nutritional and preventive application of 20 q FYM/12 q compost/ 10 q of vermicompost+5kg PSB to soil during the last tillering is deemed beneficial. Soil treatment can be done by application of 500 liters of jeevamrut/ha to the soil.

Cultivation

Pruning of the main shoot tip and secondary branch tips are done between 50-60 days of germination. This results in production of large number of tertiary shoots bearing pods. This activity increases yield by 30- 50%.

Weed management

Weed management is essential during the first 2 months of crop growth. The standard weeding schedule for red gram is 25 and 60 days after sowing. The cut weeds could be used as mulches for the crop.

Green manuring

Pre-monsoon sowing of 1-2 kg seeds of sunhemp, horse gram, cow pea, green gram, black gram and sesban are recommended. The crops can be cut down and incorporated by shallow tilling into the soil after 30 days of growth. Red gram could be sown in field one week after insertion. Incorporation of 10 q of neem leaf manure into soil has been found to reduce soil borne pathogens and nematodes.

Water management

Red gram requires very little water during its growth. Its water requirement has been calculated to be 40 cm for its entire cycle. Moisture is a critical factor during the budding, flowering and pod formations stages of red gram.

Pest management

Pod borers/boll worms (*Helicoverpa*), aphids, jassids, thrips, mites etc. are some of the major pests on red gram. Boll worm incidences is more damaging and hence of major concern. Some of the measures taken up are:

-Increasing biodiversity through planting of trees like neem, babul, pongam on the farm bunds

-Intercropping of red gram with moong, soybean, groundnut

-Random planting of marigold and hibiscus (*Hibiscus subdariffa*)

-Spraying jiggery powder (10 kg/ha) on soil surface, acts as an ant attractant. The ants predate on the larvae in soil

-Installation of bird perches to attract predatory birds

-Release of *Chrysoperla* (5000 eggs) a fortnight after sowing and *Trichogramma* (50,000 eggs) 30 days after sowing

-Spraying 5% neem seed kernel extract at 15 days interval

-Spraying of 1000 ml of HNPV (nuclear polyhydrous virus) per ha helps control caterpillars

-Spraying of garlic, chili and neem crushed in cow urine on leaves

Beetles affect red gram seeds in storage. Mixing of crushed neem leaves with seeds before storage in gunny bags is recommended. The gunny bag treatment with 5% neem oil reduces any chance of storage pest attack.

Disease management
Fusarium wilt disease is a major problem in red gram. Phyto-sanitation, crop rotation, use of resistance varieties and seed treatment are the general measures taken against diseases incidence in red gram.

Harvesting
The browning of 80% of pods and heavy shedding of leaves are the symptoms of pod maturity for harvesting.

Yield
Intercropping in rain fed areas results in average yield of 15-20 q/ha while monocropping under irrigated conditions provide 25-30 q/ha of crop.

Sugarcane (Saccharum officinarum)

Introduction
Sugarcane is a native to India. It is also grown in Brazil, Cuba, Pakistan, Thailand, Philippines, Argentina, Columbia, Indonesia and South Africa. Uttar Pradesh, Karnataka, Maharashtra, Andhra Pradesh, Tamil Nadu and Bihar are among the major sugarcane growing states in India. Sugarcane is mainly used for the production of white sugar, gur and khandsari in India. Molasses, a by-product is also used in the production of alcohol.

Climate and soil
Sugarcane is a nutrient and water demanding crop. It is not suitable to high temperature zones. For optimal productivity, it requires 750-1200 mm of rainfall during its growth period. Well drained, alluvial to medium black cotton soils are best suited for sugarcane cultivation. Cultivation in sandy loam soils under irrigated conditions also has shown some results.

Varieties
Some of the recommended varieties for organic sugarcane production are Co 8021, Co 86032, CoG 94077, Co 86249 etc.

Land preparation
Deep ploughing with disc plough followed by shallow ploughing 3-4 times using cultivar, is recommended.

Spacing
A minimum row spacing of 90 cm to maximum of 150 cm can be followed. Furrow must be prepared at 20-30 cm depth.

Organic manure
Farmyard manure of compost must be applied at the rate of 80 t/ha before ploughing or planting.

Planting material
Setts from 6-8 month old disease free nursery crop has to be selected. Always search for organically grown crop if possible.

Seed-sett treatment
Dip the setts in suspension of 2kg of azotobacter+2kg azospirillum+2kg PSB biofertilizer in 200 liters of beejamrut, for 30 minutes

Sett rate and planting
A spacing of 90 cm allows 75,000 setts to be planted in a hectare.

Green manure intercrop
Green manure crops like sunhemp, daincha can be sown, 3-4 days after planting. The green crops have to be incorporated into the soil, 45 days after its sowing. Cowpea, coriander, moong, groundnut, Bengal gram are commonly used as intercrops for sugarcane.

Crop rotation
Sugarcane is generally used in a 2-3 year crop rotation with cotton, paddy, sorghum, maize, potato, pea, wheat etc. Some of the most popular rotations are:

Maize-potato-sugarcane

Cotton-sugarcane-chickpea

Rice-groundnut-sorghum-ragi-sugarcane

Maize-wheat-sugarcane-sugarcane ratoon

Rice-sugarcane-wheat

The cultivation of legume crops prior to sugarcane is proven beneficial.

Weed management
A weed removal schedule of hand hoeing and weeding at 30, 60 and 90 days after planting is to be scheduled. Avoid chemical weed control practices.

Biofertilizers
At 30 and 60 days after planting, apply 5 kg each of *Azospirillum* and Phosphobacteria respectively. Mix the biofertilizers with 500

kg/ha farmyard manure and apply to soil. Irrigation should follow immediately.

Trashing
Dried and old leaves are to be removed at the 5th and 7th month and mulching should be done in alternate furrows.

Pest management

Early shoot borer
Trash mulching, frequent irrigation, light earthing up at every 35th day will reduce the incidence of this pest. In case of severe occurrence, release 125 fertilized female *Sturmiopsis* parasite/ha at 45-60 days of planting. Release of *Trichogramma chilonis* (50,000/ha) 45 days after planting helps in controlling the pest.

Inter node stem borer
-Use of resistant varieties like CO 975, CO 7304, COJ 46

-Phyto-sanitation

-Collection and destruction of borer eggs

-Use of pheromone traps

-Release of *Trichogramma chilonis* (2.5 ml/ha) six times from the 4th month of planting at fortnightly interval

-Use of larval parasitoids *Stenobracon deesae, Xanthopimpla nursei, Apanteles flavipes*

-Release of pupal parasitoids *Tetrastichus ayyari, Trichopilus diatraeae, Xanthopimpla stemmato*

The following methods have been found to be effectively against white flies, thrips, mealy bugs and mites:

-Spray 1:15 liters of milk-water solution, boiled and cooled

-Foliar spray of 5 days fermented whey (in copper vessel)

-Spray fermented cow urine (5 days old in copper vessel)

Disease management

Red rot disease
-Selection of resistant varieties

-Sanitation of field and incineration of diseased clumps

-Block irrigation water from infested area reaching healthy areas

-Crop rotation with rice

Smut disease
-Use of disease free stocks for setts

-Sanitation of field

-Use of resistant varieties

Grassy shoot disease
-Steam therapy of setts to 500C an hour

-Use of disease free source for setts

Cane harvest and yield
The harvest of the crop should be done when the sugar content reaches to 16% and juice purity more than 85%. The crop generally matures within a year of cropping. The canes have to be harvested 2-3 cm above the ground level. Topping should be done at the point of break.

www.ingramcontent.com/pod-product-compliance
Lightning Source LLC
Chambersburg PA
CBHW051829170526
45167CB00005B/2214